SINGLE PAYER

AND THE MERRY-GO-ROUND OF MONEY

MONEY

WHERE WE ARE

HOW WE GOT HERE

WHERE WE ARE GOING

A SUFFICIENTLY SUPERFICIAL BOOK

SINGLE PAYER

AND THE MERRY-GO-ROUND

OF MONEY

WHERE WE ARE

HOW WE GOT HERE

WHERE WE ARE GOING

TOM WEATHERS

Gastonia NC

possumgolightly press

ISBN 978-1-304-12385-5

Published by Possumgolightly Press

Ed 1 Rev 6

Gastonia, NC USA

possumgolightly.com

In conjunction with Karen Griffin of the Charlotte Health Care Justice chapter of the Physicians for a National Health Program.

Cover art by Allie Costner

CONTENTS

INTRODUCTION

Moral Imperative

This little book is based on the premise that health care should be a right not just a commodity. It should be like fire protection, police protection, clean water and other services provided by most "civilized" countries.   It should be equally available to all citizens. It is a moral imperative.

Underlying Theme

Despite the moral imperative, health care is being treated as a commodity. Not only does this create ethical dilemmas, it makes the system expensive and difficult to manage. Health care is not a classical market based system. Supply and demand functions need to be performed by a government mandated master market - typically called Single Payer - although it could be called Medicare.

Parts

The book has four parts. They are:

1.   Merry-Go-Round
2.   Single Payer
3.   Insurance
4.   History

Given that you, the reader, are likely pushed for time, the first part presents a metaphor, the second part the conclusion and the other two parts the background. You can get by just reading the metaphor and the conclusion.

Blog Version of Book

There is also a blog version of this book. The sections are arranged in a grid. The sections can be read from left to right in a linear fashion or randomly in whatever order you choose.

Sufficiently Superficial

A health care expert, someone immersed in the details of the problem could not have written this book. From the expert's point-of-view, some parts of the book will appear ridiculously simplified (or simply missing). It had to be done by a writer/analyst willing to be sufficiently superficial – someone willing to eschew nuance without too much shame.

The idea is not to present expertise but to put the facts in perspective. It is a framework for hanging the detail.

Horror Stories

Usually a book like this includes a variety of health care horror stories. Some books are structured around the stories. Being a sufficiently superficial book we don't have time. But you know the stories anyway. Somebody dies because their health insurance is dropped. A parent has to drop their own health insurance because a sick child needs it more. Somebody else pays so much for health insurance premiums that their way of life is altered – no more eating out, no more vacations, no more pre-school. The mother has to go back to work and grandmother or a friend must look after the kids.

The list goes on. But the stories are still true.

Note: Most of the stories center on insurance companies. Although not without blame they are only where the horror stories happen. Everybody is involved.

Freedom –vs – Fairness

All versions of Single Payer rely on government in one way or another to supply missing market clout and to enforce moral imperatives. Those who hate government will hate single payer

However, if anything is ever to be done about health care, the two sides need to understand each other.

There are several levels to the issue.

(Pure speculation follows)

The first level obviously is between those who hate government and those who don't. Since single payer would have to be implemented by big government it is ipso facto hateful.

The next level has to do with freedom versus fairness. As discussed in my blog of the same name (the contents of which are contained in the appendix of this document) conservatives tend to value freedom more and liberals tend to value fairness more. A conservative wants less government. He wants to be left alone to pursue her/his own interests. Liberals believe government needs to enforce fairness for all people – even if some freedom must be curtailed. Conservatives are about the individual and liberals about the community (of course this is a blanket statement with many exceptions).

The third level has to do with free will. Conservatives believe that people have the ability to look out for themselves. People with free will can pull themselves out of their own difficulties. Intrusive government restricts free will and encourages sloth and dependency. Loosen the grip of government and people will be forced to use their own free will to save themselves. Liberals, although probably claiming to believe in free will don't act that way. The implied assumption of an educated elite is that the lower classes, because of environment and other factors, are not able to help themselves. They must be helped, fixed. A conservative would argue that most liberals are busybodies.

The final difference, perhaps the most difficult for conservatives to admit and liberals to accept is that there are fundamental differences between people. Some people are simply less capable than others. Some people will always rise to the top of heap. Some will stay at the bottom. Some- most of us – will be in the middle

If this true – whether the differences are acquired or inherited – the real issue between conservatives and liberals is what to do about "these" people.

Returning to health care issues, it seems to me that conservatives must accept the basic moral imperative that regardless of their abilities people should not be allowed to die regardless of their place on the social scale. Liberals must understand that many people cannot be fixed. Some people will continue to make bad choices that cause bad health. Liberals must also realize that those on the upper end of the scale will always get better treatment than those on the lower end. And both sides must realize that ultimately all people die.

If anything is ever going to get done, the preaching will have to move beyond the choir to include the congregation.

MERRY-GO-ROUND

The Merry-Go-Round Metaphor

All financial systems are like merry-go-rounds. Consumers put money on the merry-go-round; providers take money off, serving up products and services in return. Sometimes a third party rides the merry-go-round and manages the flow of money.

The speed of the merry-go-round is determined by how much money changes hands. The wheel moves faster when consumer demand for products and services is great. The wheel moves slower when supply exceeds demand. If the wheel goes too fast there is inflation. If the wheel moves too slowly there is depression.

This analogy describes the basic capitalist market system. Generally supply and demand are balanced and speed remains steady. Cost, also reflecting the relationship between supply and demand, tends to remain within an acceptable range.

Health Care and the Merry-Go-Round

Health care can also be described by the merry-go-round metaphor – but there are two main differences, both related to cost.

1) In health care systems the third parties (insurance companies, government agencies, etc.) sitting in the middle of the merry-go-round are much more pervasive. The large sums of money that pass through the system must be managed. Money collected from recipients of care is passed to providers. A pool of money collected from healthy people is used to offset the cost of sick people. It is all very complicated. In return for providing these services, private sector middle parties (mostly insurance companies) get to siphon off some of the money as profit.

2) Unlike market based systems, health care systems cannot use supply and demand to regulate cost. Customers cannot negotiate with providers to regulate cost because true medical costs are difficult – if not impossible – to calculate. It is not like switching from a Toyota to a Hyundai because the Toyota is too expensive. Either in the name of efficiency or obfuscation, procedures and billing are scattered among a maze of entities. The customer has no one to negotiate with. Another reason that supply and demand doesn't work is that often the supply being demanded is a product or service necessary for maintaining health. In this case, there will be an attempt to pay the price – whatever it might be.

In the health care system, cost is managed by a mixed collection of for-profit entities (insurance companies and providers) and public agencies (such as Medicare). Given no single powerful market source representing consumers, cost is essentially uncontrolled.

So what does this mean in terms of the merry go round? Right now a great deal of money is flowing through the system. The wheel is moving very fast. Prices keep going up and up. An inflationary spiral threatens. However at some point, all the money will have been used up. Then the wheel will grind to a halt. There will be a depression.

What is the answer? Something else, perhaps a single payer sitting on the merry-go-round controlling the flow of money is the answer.

That is what this book is about.

SINGLE PAYER

SUMMARY-SINGLE PAYER

Health care is about life and death – health and sickness.

It is also about cost. Cost has always been a concern in health care; no service is ever quite free. However beginning in the 20th century, health care cost has become an issue rivaling health care itself. In the U.S. health care cost is greater than it should be.

The solution often offered is called "single payer". That is what this section is about.

REASONS FOR INCREASED COST

As noted in the previous section, "Merry Go Round of Money", the underlying reason for increased cost is treating health care as a for-profit commodity. In health care systems, demand has little effect on supply. Cost can go where it will.

These factors also affect rising cost.

1) As chronicled in the History section of this book, health care has changed. It has become more effective and in the process more complicated - requiring more devices and a greater number of highly trained expensive people. Large sums of money are required to operate such systems. Such increases in cost are legitimate –although some argue the salaries are too high and the equipment unnecessary.

2) Over the past 100 years or so, as health care became more expensive, more and more money passed through the system. Business saw the opportunity of getting some of this money as profit. The goal of for-profit health care systems is not only to cure people but to generate profit. Even in "non-profit" institutions, additional money can be used for salaries, facilities, creating medical monoliths by acquiring other institutions, etc.

3) Health care service and product providers can virtually charge what they will, thus increasing the money flowing through the

system and the opportunity to get more profit. As stated many times in this book, the usual rules of supply and demand don't work. Except for Medicare little control is exercised over cost. Insurance companies offer some management by restricting reimbursements. However such control can also be viewed as a way to maintain insurance company profit margins rather than a means of reducing consumer cost. Profit margins can also be controlled by eliminating high risk customers, increasing co-pays, increasing premiums, etc.

4) Another difficulty in controlling cost is figuring out what cost is. The health care system is shrouded in complexity (some say by design.) Hospitals are a maze of interconnected independent and semi-independent entities each setting its own fees. Even if customers were so inclined, there is no basis for comparison shopping. In a 200 item hospital statement the consumer cannot compare apples to apples – or even pears.

5) The emotional aspect of buying health care is also different – from buying a car for instance. Comparison shopping does not seem right. The person buying health care tends to spend whatever it takes, or at least whatever they've got.

COMPARISON WITH OTHER SYSTEMS

U.S. health care costs more and delivers less than health care in most other countries. U.S. health care costs so much that many people can't afford it. Some are dying.

A variety of statistics compare U.S. health care with the health care of other "developed" countries. Here are facts gleaned from two sources...

World Health Organization (WHO)

1. The U.S. spends more of its GDP (Gross Domestic Product) on health care than all of the other 191 countries in the survey. The U.S. ranks 37th.
2. The U.S. performs worse on life expectancy and infant mortality statistics.

3. The U.S. has the worst record of deaths that would have been preventable, given adequate early health care.
4. The typical U.S. citizen sees a doctor less often than citizens of other countries.
5. A U.S. citizen spends 5.6 times more for a night in the hospital than a Japanese citizen.
6. Although 47 million of our people are uninsured, we spend $2,797 more per patient per year than other industrialized countries.
7. The source cites a study comparing high spending areas to lower spending areas. According to the study patients in a higher spending area (Medicare comparison) had a greater risk of dying. The source speculates that this might be due to the possibility of an increased number of tests in the high spending areas. According to the Payment Types listed later the availability of more money would allow the industry to use the preferred Payment By Treatment plan, which would allow additional tests. The source wonders if possible complications of increased testing would explain the higher mortality rates.

Organization for Economic Co-operation and Development (OECD)

1. The average life expectancy in the U.S. is 78.2 years. The average in OECD countries is 79.5 years.
2. The U.S. has 2.4 physicians per 1,000 people – compared to OECD average of 3.1.
3. The U.S. had 2.6 hospital beds per 1,000 people in 2009; the OECD averaged 3.4 beds.
4. The U.S. spent 17.6 percent of GDP on health care in 2010. On average, OECD countries spent 9.5 percent.
5. The U.S. spends more on all care - especially ambulatory and administrative services than other OECD countries. For example the U.S, spends $7,910 a year versus

   $5,270 spent by Switzerland, the next highest spending
   country.
6. In the U.S. an average hospital stay is over $18,000.
  OECD countries average about $6,200 per stay.
7. Compared to other OECD countries the U.S. performs
  more tests and procedures. (Critics of these studies
  claim that differences in testing are due to differences
  in patients' needs. Studies by Dartmouth Institute do
  not bear this out.)

However, On a Positive Note

1. The U.S. leads the world in health care research. The
  FDA's comparatively short drug approval processes
  means that cutting-edge drugs and treatments are
  available more quickly to American patients.
2. The U.S has safer hospitals and better health care
  quality.
3. Because of its size and diversity the U.S. system does
  more experimentation.

COMMON REQUIREMENTS

Across the world, all successful health care systems satisfy these
basic requirements.

Cover all people

Satisfying this requirement tends to render health care systems
less complicated and easier (and cheaper) to manage.
Compared to other countries the U.S. spends much more on
administration. Further, by including all people under the plan,
risk and cost are spread over the largest possible pool of
participants. (At the time of this writing, one criticism of
President Obama's Affordable Health Care Act is that it is too
easy for people to opt out of plan, thus reducing the pool
needed to support the system. See "Side Bar Topics" later in this
section for a summary of Obamacare.) Covering all people also
satisfies the moral imperative that everybody has equal access
to health care.

Reduce the role of insurance companies

As profit-based amoral entities it makes sense for insurance companies to do anything more-or-less legal to minimize cost and maximize profit. One thing they can do is limit coverage of the most risky population – the aged and the sick. Weeding out expensive people requires expensive administrative systems whose cost is passed on to consumers. Further by handling all money passing through the system these amoral entities have the power to extract profit from every transaction.

As powerful as insurance companies seem to be, their power is limited. There are so many companies competing for market share that no single entity can negotiate cost. For the most part health care providers can charge what they will. Insurance companies, squeezed by providers, maintain their profits by shifting the pressure down the line to consumers who must ultimately bear the burden of a dysfunctional system.

As demonstrated in other countries, changing or getting rid of insurance companies is a major means for reducing costs.

Recognize health care system for what it is

As we have noted many times, a health care system should not be treated as a for-profit enterprise. It is more akin to public institutions such as the police or fire department. As standard supply and demand controls did not work for police and fire departments (they were tried) such controls will not work in health care systems

Like a police or fire department, the basic solution requires a public entity powerful enough to implement the requirements of the system – one of which is to provide services for everybody. That powerful entity is our representative government (local for fire and police departments, national for health care systems).

Funneling payments through a broad-base master-market (e.g., a "single payer") would generate the clout needed to manage costs. Single payer would supply the missing cost control function. Such a system, being aware of matters of life and death would also ensure that the moral imperative is satisfied. Everybody would be covered.

SINGLE PAYER

"Single Payer" is a term often used to describe a master-market function that serves public good rather than private profit. Single Payer makes the Merry-Go-Round of money work for health care. Responding to the basic moral imperative Single Payer ensures that everyone has equal access to health care.

Note: Medicare Master-Market

Medicare is a single payer master market system negotiating with providers to reduce costs and ensure public good. Medicare is an example of the National Health Insurance (Canadian) variation of Single Payer.

Many Single Payer proponents prefer the term Medicare. It is thought to be more politically acceptable. Proposed single payer legislation (HR 676) is called " Expanded and Improved Medicare for All".

Single Payer Variations

There are almost as many variations in the Single Payer approach as there are Single Payer plans. Some do not even have single payers. Here, based on T.R. Reid's "The Healing of America" are the basic variations:

Bismark Model

This system is employed by Germany, Japan, Belgium, Switzerland and some other countries that might be regarded as fans of Über military man Prussian chancellor Otto von Bismark who invented the model. Not strictly a single payer system (Germany has more than 200 funds), this model uses private providers who receive money from private non-profit insurance companies. Payment comes from employee payroll deductions.

The U.S. follows the Bismark model with one major difference. In the Bismark model insurers are non-profit entities. Not so in private U.S. health care systems.

Beveridge Model

This system, employed by Britain (where it was invented), Italy, Spain, Cuba, and most of Scandinavia comes closest to what is generally termed socialized medicine. People pay a health tax; the government pays for medical care. There is no direct fee. Most providers work for the government. Two examples of the Beveridge Model are Cuba and the U.S. Department of Veteran's Affairs.

National Health Insurance (Canadian) Model

This system employs private providers and a single government-operated insurance plan. The plan collects money from all citizens and pays bills. Money is saved by (1) expanding the pool of people covered, (3) reducing complexity (3) eliminating insurance company profits and eliminating the complex systems used by insurance companies to classify customers and managing claims, and (4) being able to negotiate with providers and drug companies to reduce cost. The government, by taking control of money away from insurance companies, has the power to exercise missing market clout – which ensures that the moral imperative is observed. This system is used by Canada, who developed it, Australia, South Korea and other newly industrialized countries. As noted before, it is also used by the U.S. Medicare system.

When Insurance Is Not Used – Out-of-Pocket Model

With the out-of-pocket model, you pay what you have. The rich get care and the poor either die or if they are lucky get charity. This is the model used by most third world countries and by uninsured U.S citizens.

Plans Used in the U.S.

In the U.S. a patient's health care depends on who the person is (although the analogs with other systems are not entirely perfect)...

- Workers under 65 are covered under a variation of the Bismark plan (but without the same controls and clout – for-profit insurance companies drive the U.S. variation).
- Retirees over 65 are covered under Medicare which is basically a National Health Insurance style (Canadian) system.
- Native Americans and military personnel are covered under a Beveridge style system (e.g., socialized medicine).
- Uninsured Americans are covered under the out-of-pocket model. Unless they have money or access to a charity, a serious health problem is a death sentence.

SIDE BAR TOPICS

These are important topics that don't quite fit in the simplified structure of this document.

"Obamacare" (The Patient Protection and Affordable Care Act – PPACA)

PPACA has been termed "hideously complicated". There is no way it can be completely explained in this little book. We will have to stick with my "sufficiently superficial" model of authoring. I shall eschew nuance.

(Be patient regarding complicating exceptions. I'll get there.)

The purpose of Obamacare is to ensure that everyone has health insurance. Right now most people get insurance where they work.

Under this new system, much insurance still comes from employers. However those not covered by employers must provide it themselves.

Money for people who get their own insurance comes from tax credits. The amount of credit is based on income. Those with more income receive less credit; those with the least income receive more money. People with no taxable income can get money from Medicaid, some of which is funded by the federal government, some by the states.

People get their insurance through "exchanges". Each state has its own exchange. The exchanges list plans offered by participating insurance company. All plans offer the same basic coverage. Plans are organized in four tiers - Platinum, Gold, Silver, and Bronze. Platinum offer the best benefits, Bronze the basic coverage. All plans offer choices between lower deductibles and higher premiums.

(Complications)

Companies with over 50 fulltime employees must offer insurance. Failure to do so will result in a fine. COMPLICATIONS: A company can reduce employee hours so they do not count as full-time employees. If the fine is less than the cost of insurance it might make more sense for the company to not provide the insurance. Smaller companies not now offering health would see their insurance cost go up. They might, as noted above, reduce staffing below 50.

Individuals will also pay a fine if they do not get insurance. COMPLICATION: If the fine is less than the insurance premium, younger healthy people might not get insurance. A pool of healthy people paying premiums is needed to pay for sick people. If enough people do not get insurance, there might not be enough profit to make it worthwhile to provide insurance.

People who have been receiving untaxed income under the cover will run into problems when filing their first returns. COMPLICATION: Such people might opt of the new system.

States are expected to supply money for the new system. Conservative states might refuse to supply the money. COMPLICATION: There might not be enough money to survive.

Payment Types

The present system employs various payment types – some that are less costly and more efficient than others. Following are basic payment models with the lowest and highest cost systems indicated.

Payment for Performance – Lowest Cost

Payment is based at least in part on how well treatment works. The information on which payment is based can include: (1) whether activities associated with successful outcomes have been performed (2) whether the outcome is deemed successful by given criteria (3) how well patients regard their treatment (4) whether factors not associated with a particular patient – such as overall low rates of infection – are present.

Payment for Treatment – Highest Cost

Providers are reimbursed for each treatment episode. Payment is based on the quantity of treatment rather than the quality. Given the tendency of providers and patients to ensure that every diagnostic possibility is reviewed (including those demonstrably unlikely to bear on current case) multiple tests are often employed.

Payment by Diagnostic Group (DRG)

Hospitals combine diseases into groups according to the resources needed for care, arranged by diagnostic category. A dollar value is assigned to each group as the basis of payment for all cases in that group, without regard to the actual cost of care or duration of hospitalization of any individual case. By separating the quantity of treatment from the actual treatment, cost is said to be reduced.

Retainer (also called capitation)

A provider receives a retainer based on the number of patients expected to be seen for a given period. This type of payment system is common in HMOs (health maintenance organizations). PPOs (preferred provider organization) typically pay per treatment or visit.

Complexity

Health care is an increasingly complex system. Today the complexity of health care management rivals the complexity of health care itself. A major benefit cited for single payer is at least some simplification – which makes the system more efficient and reduces costs.

The rise in complexity can be regarded as an ever rising curve variously called asymptotic, Pareto and geometric. The main feature of such curves is that over time they get steeper and steeper. There is the long period when the curve remains nearly flat then a sudden rise when the curve gets infinitely close to but never reaches the vertical. Before that impossible moment arrives, the complexity that drives the ascent becomes too great, too fragile to be sustained.

Joseph Tainter in his book Collapse of Complex Societies says the result is likely to be a collapse. The appendix includes a section from my book "Crossing Infinity" that summarizes some of Tainter's ideas.

Prevention

Costs can be reduced by improving people's health thus reducing the need for health care. Everybody knows what to do – lose weight, get more exercise, stop smoking, get regular checkups, etc. Earnest well-meaning people become health evangelists – giving lectures, writing articles. Of course this is laudable. However, one wonders - how well will people who walk because they don't have cars relate to those living in neighborhoods where people walk and run for health.

And of course the Live Free or Die crowd and all others in opposition to anything regarded as progressive will simply hate the advice.

Consolidation

Businesses sometimes consolidate – smaller companies combine into bigger companies. It is an accepted practice. Economies of scale allow a bigger business to do things cheaper and thus be able to charge less for their products – and thereby become more competitive and attract more customers. In a normal supply and demand economy this makes sense for the consumer and for the business. (Of course it doesn't make sense for workers who lose their jobs as a result of improved efficiency brought about by consolidation.)

One would think the purchase of smaller private practices by hospitals would also benefit both the business and the consumer – thus reducing costs and attracting more business. In terms of information sharing, consolidation has been a benefit. Institution-wide computer systems allow information gathered in one location to be immediately available at all locations.

However the costs charged by the former private practices have gone up not down. One reason is that Medicare pays more for the same procedure performed by a hospital than when performed by a private practice. After becoming part of the hospital, the procedure can be charged at the same rate – even though the underlying costs have not changed.

Another difference between consolidation in health care and other businesses is the handling of duplicated procedures and equipment. In a usual business, consolidation allows the elimination of duplication and thereby an increase in efficiency (and a reduction in costs). That is a major goal of the consolidation. However this process does not hold true for health care. The former private practice tends to duplicate many procedures performed somewhere else.

Of course one could argue that retaining local-level operations is more convenient for "customers" and broadens the base for the larger corporation. However it could also be argued that in a normal competitive environment, systems with lower cost would attract more customers and be more likely to prosper.

This process bears out one of the one of the underlying premises of this book – that the normal procedures of supply and demand do not work in the health care business. Handling this situation requires the clout of a Master-Market.

.

INSURANCE

SUMMARY - INSURANCE

This section is about fees and payments (and perhaps villainy).

Prior to the early 20th century - since prehistory really - medical payments were mostly on a sliding scale. People paid what they could afford. For instance, one source claims that in ancient Egypt the well-to-do might pay 10 shekels for treatment whereas poor people paid two shekels.

The sliding scale ended when insurance companies insisted on a standardized fee system. (One supposes anything else would have been an actuarial nightmare.) Oddly, perversely, sliding scales are still around today, except that now those without health insurance are charged the most.

Also there is the issue of medical commercialization. At various times in Europe, doctor's fees were restricted. However, in America, with its emphasis on freedom and individual autonomy (take this gun from my cold dead hands, etc.); medical practice was always regarded as a commercial venture. The sliding fee system was in part a response to economic realities – it was simply how one got paid. Another better way is OK.

The coming of medical insurance – and the need for it – changed all that. Ever since the 1930s, when health care became more effective and thus more expensive, insurance companies have been funneling payments through an insurance system of one sort or another. They collect money from a large pool of healthy people to pay the medical expenses of a smaller group of sick people

After the aside below, this section discusses major steps in the evolution of medical fee payments. There are passing references to fee systems in the eighteen and nineteen hundreds. However most of the activities take place in the 20th century.

ASIDE - VILLAINS

Insurance companies are sometimes regarded as the villains in this story. It is not so simple.

Consider...

Insurance companies are "amoral".

This does not mean that insurance companies are "immoral" – that they deliberately do bad things. It simply means they don't seem to care. Morality is not the issue; it is profit. Of course every capitalist enterprise is amoral to one degree or another. However in most other business, greed is regulated by the market itself, by government, or simply by the moral sense of the CEO. Given the evidence, this does not seem true for insurance companies. In the pursuit of profit, the sick and the elderly are regularly culled from the roles of the insured. People die so that insurance companies can make more money. (I am sure that people who work for insurance companies would protest that they are not amoral people. I would suggest that these people are isolated by layers of bureaucracy from the consequences of their actions – in the same manner that a bomber pilot is isolated by miles of air from the destruction he causes.)

A shift in perspective is required when viewing insurance companies.

Insurance companies did not come into existence to ensure that patients got medical care but to ensure that health providers got paid. See "Up to Modern Insurance".

Insurance companies are not all powerful.

Given the reputation of insurance companies it is easy to imagine that they are all powerful. That is not true. Although insurance companies do influence cost to some degree, in the end it is the health care providers (hospitals, equipment manufacturers, pharmaceuticals) who set costs. As a collection of disparate entities individual insurance companies simply do not have the negotiating clout. That would take the power of the master market described in the previous section.

CHRONICLE

1 - Up to Modern Insurance

1850 - Franklin Health Assurance Company of Massachusetts provided accident insurance to cover injuries related to railroad and steamboat travel.

1800s and early 1900s – Prior to the late 19th and early 20th centuries, people generally didn't seek medical care. It wasn't that effective. Most insurance was disability coverage to ensure income continuation. However medicine advanced and the public image of medicine changed. New techniques and devices were developed. Medical schools were accredited

With the rise in regulations and quality of health care, demand for medical services increased faster than the supply of physicians and hospitals. The combination of these factors brought an increase in medical costs, which prompted the development of modern day health insurance

1929 - Dr. Justin Ford Kimball, an administrator at Baylor University Hospital in Dallas, Texas, realized that many schoolteachers were not paying their medical bills. In response to this problem, he developed the Baylor Plan – teachers paid 50 cents per month in exchange for the guarantee that they could receive medical services for up to 21 days of any one year. The purpose of the insurance was not so much to improve patient care but to ensure that the hospital got paid. This is important.

2 - The First Non-Profits - Blue Cross and Blue Shield

Blue Cross

1930s – In response to financial pressures of the Great Depression more hospitals adopted the Baylor Plan. Medical insurance became much more widespread. Such health plans enabled consumers to be insured but also gave hospitals a steady income despite economic turmoil. However, these single-hospital plans also generated price competition. To avoid the loss of income that might result, community hospitals worked together to create health coverage plans. (Note that insurance companies were not serving the needs of the insured but the providers. Also note that great effort was expended to ensure that normal supply/demand controls did not apply – a major theme of this book.)

1939 - The American Hospital Association (AHA) first used the name Blue Cross to designate health care plans that met their standards.

1960 - The independent Blue Cross plans merged to form Blue Cross under the AHA in 1960. Considered nonprofit organizations, the Blue Cross plans were exempt from taxes, enabling them to maintain low premiums.

Blue Shield

1939 – Fearing the loss of autonomy and to ensure their interests were protected plans covering physician and surgeon services emerged around this time. Again note – the plan was provider oriented not patient oriented.

1946 - Physician-sponsored plans combined into Blue Shield.

Merger

1971 - Blue Cross and Blue Shield merged into one company.

Fees

Initially Blue Cross and Blue Shield charged the same premiums for all customers.

3 - Rise of For-Profit Insurance Companies

1930s - Several life insurance companies began to offer health insurance. Premiums were charged according to calculations of relative risk, charging more money for older people and for people with a history of a medical condition. Blue Cross and Blue Shield Plans, previously in the nonprofit sector, were forced to compete with commercial health insurance companies, and eventually began to charge premiums in the same way – by basing payments on risk assessments.

4 - Employee Benefit Plans

1929 –The Ross-Loos Medical Group was established to provide prepaid health insurance for businesses. It is considered to be the first HMO (Health Maintenance Organization) style insurance in the United States. Headquartered in Los Angeles it had approximately 500 members enrolled at a cost of $1.50 each per month.

Note: The terms HMO and prepaid health plan are used somewhat synonymously, although "HMO" has become more common. The term "cooperative health plan" has also been used.

1940s and 1950s - Employee benefit plans became more and more comprehensive as strong unions negotiated for additional benefits. These benefits included health insurance. This was especially true during the Second World War. Companies competing for labor had limited ability to use wages to attract employees due to wartime wage controls. They competed through health insurance packages. (Those working for Henry J. Kaiser were the principal recipients of early HMO style coverage.) The companies' healthcare expenses were exempted from income tax. At the time of this writing, HMOs are the main supplier of workplace health insurance.

1970s – 1980s Issues with health care system and ever increasing cost of medical technology lead to increased costs for the prepaid plans then in effect. The number of such plans dwindled. However Richard Nixon's HMO act of 1973 required employers with more than 25 employees to have government subsidized HMOs. The number of prepaid plans resurged. (The HMO act amended the Public Health Service act of 1944.)

During this same period PPO (Preferred Provider Organization) were created as a cost saving alternative to HMOs. In an HMO the insured are restricted to providers who are part of a single network. In a PPO, the insured can go outside the network, but coverage will be reduced.

In previous years such prepaid plans (HMOs, PPOs) were originally nonprofit, but in the 70's and 80's the plans, were largely replaced by commercial interests.

(Some sources say that many under-the-table and behind-the-scenes techniques were used throughout these processes. The sources claim that for-profit HMOs made money by hiding costs and profits within a maze of complexity. They also claim that a huge and expensive bureaucracy was established to hold down the cost of payments. It is said that money was made in the conversion from non-profit to for-profit status.)

5 - Government Involvement

1798 - The Public Health Service Act authorized marine hospitals for the care of American merchant seamen.

1930's – Under President Roosevelt the Democrat-dominated congress passed the Social Security Act – although it did not include health insurance.

1944 – The Public Health Service Act consolidated and revised most legislation relating to the Public Health Service. The mission of the Public Health Service, amended many times, is to promote the protection and advancement of the Nation's physical and mental health.

1940's – 1950's President Roosevelt proposed an Economic Bill of Rights. It did not get anywhere. President Truman proposed a single national health care plan that would have included all Americans. The plan was denounced by the American Medical Association (AMA). It was called a Communist plot by a House subcommittee.

1954 – Under President Eisenhower, Social Security coverage was amended to include disability benefits

1965 – President Johnson pushed Medicare and Medicaid pro-grams through a Democratic majority in Congress.

1970s - President Nixon's HMO plan of 1973 required that companies with more than 25 employees have an HMO plan requiring federal endorsement, certification, and assistance. HMO's were widely adopted by employee insurance plans.

His "War on Cancer" centralized research at the Public Health Service National Institute of Health (NIH).

1980s - President Reagan shifted Medicare payment to diagnosis related groups (DRG) rather than payment by treatment. This method was adopted by private insurance companies.

1993s – President Clinton's universal health care proposal failed – partly due to his personal peccadillos

1996 – The relatively modest Mental Health Parity Act and the Health Insurance Portability and Accountability Act were passed.

5a - Government Involvement – Affordable Care Act

2010 - President Obama's Affordable Care Act. This is not universal health care, nor single payer. The plan remains hideously complicated. It has been ridiculed as "Obama Care." The haters hate it. However it does substantially expand health care coverage although there will gaps in the system.

These are the features...

Raise the Medicaid maximum income for people who do not have insurance. As before, the Federal Government and State Governments split Medicaid costs. It is not clear if states will continue to do that.

Expand employer insurance. Companies with more than 50 people must offer an insurance plan to all employers. A tax break will be offered to smaller companies.

Require that people who do not now have private insurance obtain it. States will be required to set up insurance "exchanges" where the uninsured can select from a range of plans. Federal aid will be provided to help the insured make payments.

Regulate insurance companies. The law will require that insurance companies offer coverage to everyone. Practices such as refusing insurance to people with pre-existing conditions will not be allowed.

The plan will be funded through higher taxes on wealthier people.

HISTORY

SUMMARY OF HISTORY

This is a brief (and necessarily superficial) review of health care for the past 28,000 years or so. The purpose of the review is to show how health care over the millennia has become more complex. This process started slowly then proceeded at an ever increasing rate. In the last 100 years the cost and complexity of health care has become an issue almost on par with medicine itself. Cost and complexity are what this book is about.

This section is about the history of health care and the history of health care systems. Although the two are often viewed as the same thing, they really aren't.

Health care is what medical practitioners (doctors, hospitals, clinics, drug companies, medical researchers, etc.) do. Health care happens when practitioners examine sick people, discover the causes of diseases and injuries, and use this information to help cure medical problems.

Health care systems are the relationships between the health care parts (material and immaterial). Many, but not all of modern health care problems are really systems issues. Single Payer is a systems issue – it does not require new treatment but new management.

The italicized text below provides a systems-oriented overview of each period. Even when there are few medical developments in a given period, there are always system developments. In both cases, the development in one period sets the stage for the next period.

Note: The historical facts (but not the systems analysis) in this section come largely from the web-based Schlumberger Excellence in Education Development (SEED). However, be aware that only words (e.g., doctor, drug, treatment, etc.) from this copyrighted source are used, not the sentences. Using the former is research (not using the words would be impossible);

using the latter is plagiarism. This section distills the SEED material, extracting and reorganizing facts.

PREHISTORIC SYSTEMS

28,000 BCE

A system (at least an intuition) was needed to identify future shamans and to train and pass knowledge to acolytes. An organization – at least a systematic repeatable ritual – had to be in place.

A shaman might have performed (still perform in some societies) basic medicine – treating wounds, setting broken bones, administering healing herbs, etc. (which was probably as effective as much treatment performed until recent times). However, in the shaman's world, the most significant healing came from contact with spirits inhabiting rocks, plans, water, animals, etc. In such societies spiritual and physical health are indistinguishable. (Some say that the holistic view represented by the shaman's world is superior to the systems view presented in this section.)

HEALTH CARE SYSTEMS IN EGYPT AT THE BEGINNING OF THE HISTORIC ERA

3200 BCE

The Egyptians invented writing (pictographic writing anyway). With this tool they created medical histories and documented medical practices. Writing tools are always systematic; in some cases so is the reality represented by the writing. In ancient Egypt the written material described the healing work done and when it was done (the history of the work). This process (symbols for representing information, and the reality corresponding to the symbols) is the basis for all information systems. So, viewed in one way, 5,000 years ago the Egyptians started health care on the road to complexity.

Modern investigation of Egyptian papyrus documents reveals a society with a surprisingly developed medical system.

Drugs derived from plants and naturally occurring materials were employed. Castor oil was used as a laxative. Willow tree products (containing the active ingredients of aspirin) were used to speed healing. Medical procedures included removal of cysts, male circumcision, bone-setting, and using pressure to stop bleeding. Mummifying bodies made possible the study of organs. The functions of many major organs was understood

HEALTH CARE SYSTEMS IN CLASSIC GREECE

400 BCE

Aristotle Socrates and Plato (the big three) developed the logical tools for organizing descriptions of reality (including health care systems and standards) – specifying what belongs with what, what precedes what, etc.

Many modern ideas about medicine and the functioning of the human body originated in Greece. For example diseases were thought to have external causes originating from the environment as well as internal causes. Physiology was studied by systematic dissections of the human body.

In 460 BCE Hippocrates and his students wrote 70 books that describe the medicine and health care systems of the time. The Hippocratic Oath describes medical standards still in effect today. Treatments were kept simple and the admonition "first do no harm" was an ideal. The conflict between performing medicine for public good instead of private gain was examined. That issue still resonates today and is one of basic arguments of the Single Payer issue.

HEALTH CARE SYSTEMS IN ROMAN ERA

0 CE

The Romans did not greatly advance medicine. Most Roman physicians came from Greece. However the Romans did develop practical health care management systems. For example hospital systems were commonly found in military camps located throughout the empire to allow soldiers to be treated . If the Greeks were health care practitioners the Romans were systems analysts.

Gaen, the most influential physician of the Roman era, wrote and organized 500 books on medicine, many designed to teach medical arts to new practitioners. He stressed that the best way to learn about health and disease was through the dissection of animals (not humans – that was illegal) and the study of anatomy. His works were used as standard medical references until the end of the Middle Ages.

MEDIEVAL HEALTH CARE SYSTEMS

1100 CE

Hospitals and medical schools were staffed by people organized by systems of roles and responsibilities.

As in Rome, medical practice advanced little in Europe during the Middle Ages. Medicine was not much different than in Greek times. Scholarship was primarily a religious function; clerics were more interested in saving souls than curing bodies. Disease and injury were generally attributed to supernatural causes. Cures were only possible through prayer. No new medical research was conducted, and no new practices were created.

However, there were advances in medical systems

Monasteries operated many large hospitals. The cures were not effective but the patients were generally well fed and comforted.

Medical specialization was established - physicians worked on internal issues, physicians on external (visible) problems. Other roles, comparable to modern specialties were also defined.

Medicine became a profession, requiring education and certification of sorts (although "leeches" and female practitioners for female patients were allowed). Medical schools were established based on Greek and Arabic medical sources (translated into Latin).

RENAISSANCE HEALTH CARE SYSTEMS

1500 CE

The scientific method (a tool for exploring facts and relationships in physical systems) and the objective attitudes it instilled became the basis of research.

Although medical practice only evolved modestly from the previous period, science moved beyond traditional views that governed medicine in both the East and the West. Treatment was no longer derived from a divinely ordained natural balance.

The greatest contribution to research in all areas of science was the adoption of the scientific method. The scientific method is a technique for investigating problems – a hypothesis is stated, experiments are performed, data is collected and examined, and finally a theory based on the observations is proposed. The process can be repeated by any objective researcher and the theory proven or disproven.

Roger Bacon (1214–1294), developed the philosophical basis of the modern scientific method. Galileo (1564–1642) put the theory into practice.

HEALTH CARE SYSTEMS IN AGES OF ENLIGHTENMENT, OF REASON AND THE SCIENTIFIC REVOLUTION

1600 – 1800 CE

Ever more complex systems were required to manage the relationship between increasingly interrelated medical activities and knowledge. The habit of systems thinking was becoming ingrained.

The Age of Enlightenment is sometimes called the Age of Reason. Building on the ideas of the Renaissance, the Ages of Reason and Enlightenment laid the foundation for the Scientific Revolution. The ideas still argued today were first formulated by such philosophers as Spinoza, Kant, Hume, and Rousseau. The scientist Newton developed theories that would change the world. Enlightenment ideas influenced Benjamin Franklin and Thomas Jefferson during the American Revolution. Liberal and conservative disagreements stated in the Age of Enlightenment are echoed today in "Single Payer" discussions. (The topic "Freedom vs Fairness" in the appendix of this little book is a continuation of that argument.)

Medical practice greatly advanced during the 17th and 18th centuries. New knowledge abounded.

 Citing only a few examples...

Professional societies were formed in all major European capitals, and scientists shared their research by publishing in journals.

William Harvey explained blood circulation.

Using a microscope, Anton van Leeuwenhoek discovered red blood cells, bacteria, and protozoa.

Clinical practice was revolutionized by Thomas Sydenham. Detailed observation and record keeping was emphasized.

Vitamins were discovered. New drugs were created, and modern-style vaccination was performed.

All this activity required managing systems. It could not randomly happen.

LEADING TO MODERN HEALTH CARE SYSTEMS

1850 – Today

If plotted on a graph the advances in health systems would occur on an ever rising curve. Ever more complex systems have been developed to manage ever more complex institutions that are evolving to deliver ever more complex health care. Some of these systems organize money.

In the mid to late part of the 19th century modern health care began to take shape. Although, the advanced technologies and medical practices of the 20 the century were yet to developed and much basic information (e.g. DNA) was not yet known – the fundamentals were there. A doctor then could probably understand the basics of modern medicine now and a doctor today could probably understand the basics of medicine then.

As all this happened, health care systems advanced to manage what was going on. However there was one major piece of the pie missing. That was how to pay for increasingly expensive medicine and what sort of systems would be required to manage that activity.

Hence this book.

.

APPENDIXES

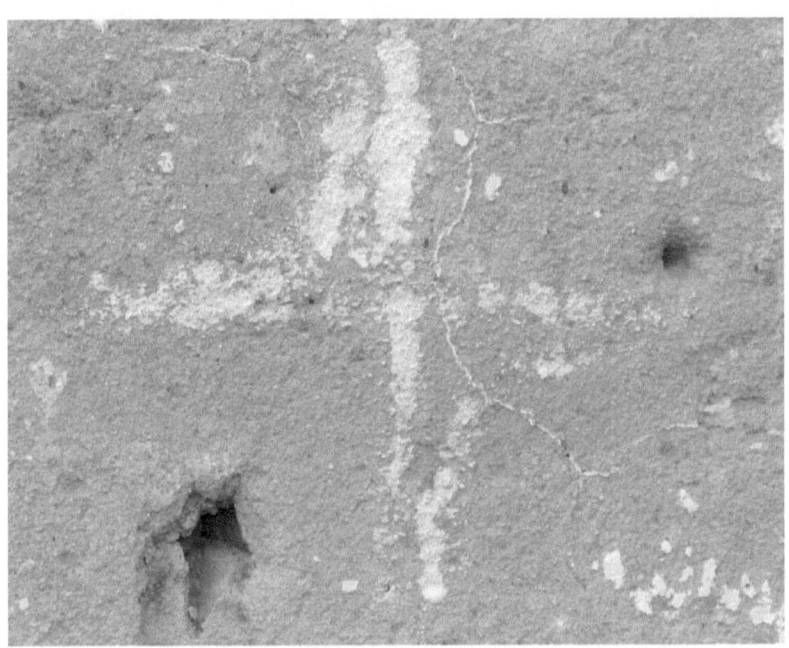

NOTE: MASTER MONEY MARKET

Throughout this book I refer to the term "master market" - or sometimes "master money market". This is an entity, most likely the federal government, that makes sure pricing remains balanced in the health care system. In a pure commodity-based capitalist system price is regulated by the push and pull of supply and demand. Such supply and demand functions do not work well in the medical system. Prices charged by providers of health care products and services are not subject to classical market pressures. A master market with the necessary clout (Medicare for example) provides this control.

Naively I thought I invented the term master market and that I was alone in attributing increasing cost to the providers of medical products and services . Insurance companies, I speculated are not the only villains. In turns out that others believe providers not insurance companies are driving up medical cost. Eraza Klein in a Bloomberg.com column argues this point. Karen Garloc in the Charlotte Observer provides the example of the $89,000 hospital bill to treat a snake bite.

INSURANCE GAMBLE

Health insurance (all insurance) is a gamble. The insurer gambles that the amount paid in by all the insured exceeds the amount paid out in all claims. The difference is profit. For profit-based systems, supply and demand (and actuarial tables) generally control premiums, although games are often played. Profit-based insurance companies make profit any way they can, pushing to the edge (or crossing over) legal limits. They hide advantage in complexity. For-profit insurance companies know no moral imperatives. Critics who rail at this reality are missing the point.

For non-profit systems, it is pretty much the same, except profit is supposed to be eliminated. The problem with the modern hodgepodge healthcare system is that no one seems to know what to charge and/or pay. True costs are hidden within arcane billing systems in which the amount charged proceeds through various iterations in ways not understood by those for whom the care was provided. Not only do such systems confuse the insured, they increase cost. That is what Single Payer is about.

RISE INTO COMPLEXITY

Health care – including Single Payer - is an increasingly complex system. Today the complexity of health care management rivals the complexity of health care itself. The rise in complexity can be regarded as an ever rising curve variously called asymptotic, Pareto, geometric and so on. The main feature of these curves is the long period when the curve modestly climbs then the sudden rise when the curve gets infinitely close to but never reaches the vertical. Before that impossible moment arrives, the complexity that drives the ascent becomes too great, too fragile to be sustained.

Joseph Tainter in his book Collapse of Complex Societies says the result is likely to be a collapse. Following is the discussion of Tainter from my book Crossing Infinity.

JOSEPH TAINTER AND COMPLEXITY

By adapting current systems to new data (e.g., adding epicycles within epicycles) we increase the complexity of the current systems.

Joseph Tainter is one of the founders of the view of societies (in our case the medical establishment) as complex systems. He described his theories in the 1988 book *The Collapse of Complex Societies*. His basic idea is that societies become complex as they adapt and change to solve problems.

For a time this works, depending on how adaptable and creative the society is. The society grows, prospers. New energy is added to the system.

However, developing and maintaining complex solutions requires additional layers of description and control. These layers cost – which means that more and more energy is required. Solutions become increasingly convoluted, consuming ever more energy. (Considering a household as a thermodynamic system, over the years more and more energy is required to implement the "solutions" needed to maintain the household. Or consider a large software project. Over time the basic programming team is expanded to include project managers, business analysts, technical writers, quality assurance testers, marketing people, etc.)

Up to a point complex solutions might increase the amount of energy available. Past that point the solutions consume more energy than they produce. The society goes into the hole and unless it stops whatever it is doing must ultimately fall.

According to Tainter, societies stick with complexity because short-term solutions tend to acquire momentum. The costs add up.

He also notes that there are more "concatenating" problems. These are issues (and solutions) that tend to interact in unexpected ways.

FREEDOM VS FAIRNESS

The most vociferous opponents of health care control are really talking about freedom – as they understand it. In order to get anything done we must respect and understand these people. When pushed, they dig in.

Following is a discussion of this topic from my blog, Freedom – vs- Fairness.

THEORY

Arguments between the conservative-right and the liberal-left (at least in the United States) can be explained by differing emphases on Freedom and Fairness.

Freedom

The conservative-right believes everybody should be free to pursue power (money, fame, health care, etc.) with as little interference as possible from big government. How power is ultimately distributed depends on the individuals. Some will always end up with more power (money, fame, health care, etc.) than others. Inequities happen. But one should still respect the authority of individuals at the top.

Fairness

The liberal-left believes that no one individual should be allowed to have undue power over another individual. Inequities should be minimized. The power of central government shall be used to restrict the power of some to protect the power of others.

COROLLARIES

Given the differing emphases on freedom -vs- fairness, other differences follow:

Free-Will and Responsibility

To the conservative-right, a corollary of freedom is free-will. Individuals know their own will and should be free to choose what is best (unless acting freely violates a moral code). People are responsible for their choices. There are no reasons for bad behavior, just excuses. According to the liberal-left, free-will is a relative commodity. Some are more free than others. There are reasons for bad behavior - extenuating circumstances. Individuals don't always know their own best interests. Those with more free will sometimes need to help, guide (or manage) those with less free will. Although people might always be held accountable for their acts, they are not always responsible.

Competition and Cooperation

According to the conservative-right, freely-acting individuals will inevitably bump into one another. The resulting competition is the basis for most human activities - politics, sports, economics. This is the most efficient way - the natural way, for getting things done. Anything else is unnatural. Competition is tough, manly. Cooperation is weak, girly. According to the liberal-left, competition must be tempered by cooperation. As a social species, it is also natural for humans to get along. Unchecked competition is primitive, testosterone diseased. Cooperation is wise.

Empathy and Ambiguity

In the pursuit of fairness and cooperation, the liberal-left is compelled to see all points of view. It is no accident that a "liberal" is writing this blog that purports to see both sides of issues. The liberal develops a tolerance for ambiguity. For the conservative-right, concerned more with personal freedom and competition, there is less need for understanding. Excessive empathy might be regarded as a disadvantage. Tolerance for chaos and ambiguity might also be less well-developed.

Unfairness and Loss of Freedom

In the pursuit of fairness for all, the liberal-left is unfair to some (e.g., graduated taxes, business regulations, trade rules, etc). In the pursuit of freedom for all, the conservative-right allows individuals at the bottom of the heap to be dominated by those at the top. Only freedom of opportunity is equal.

Both sides impose order (and sacrifice freedom and fairness) in the name of correctness. For example, the conservative-right might censor speech and freedom in order to preserve a particular hierarchical structure (resulting from some individual's exercise of freedom). The liberal-left might censor speech in order to ensure that one group does not speak unfairly about another group (this is "political correctness").

At the extremes, both the liberal-left and the conservative-right can (and have) resulted in totalitarianism. The unfettered liberal-left tends toward communism. The unfettered conservative-right tends toward dictatorship.

Correctness

There are correct expressions, utterances, views. Both the liberal-left and the conservative-right have notions of correctness that arise (or not) from their core positions. For the conservative-right, religion is often viewed as correct because it stems from individuals in the pursuit of freedom. Gun ownership is correct for the same reason. The liberal-left is concerned with speech; it should be fair and correct. Unfair speech is incorrect. Certain community aesthetics are also subject to correctness (public art, landscapes) imposed by either the right or the left - depending on who is in power.

Rights

Rights are guarantees provided by government (or God) ensuring certain human freedoms - in the US, the freedom to bear arms (more or less), freedom of the press, freedom of/from religion, etc. Although all rights generally restrict the power of government over individuals, some rights are associated more with the liberal-left and some with the conservative-right. Liberals are generally associated with rights that promote public fairness - the freedom of the press and the freedom from religion. Conservatives are generally associated with private freedoms, such the freedom to bear arms, freedom of religion, and the freedom to do what you want with your own property.

Celebrations of the Right

The conservative-right with its emphasis on individual power favors hierarchical organizations headed by dominant individuals. It celebrates the strong man, the tribe, the team. The conservative-right loves competition, aggression, dominance. It respects authority (of the right sort).

Hubris of the Left

Symbols of the Right When devising rules of fairness for the operation of a political system, the liberal-left must presume to understand the operation of the system. Those devising the rules of the liberal-left must assume they know better than those for whom the rules are being devised. When exercising power in the name of a public esthetic the liberal-left must assume that its esthetic is correct.

- Mel Gibson shouting "Freedom" in Braveheart.
- Charlton Heston holding a musket in front of the NRA giving his "cold dead hands" speech.
- A flag (especially the Confederate battle flag).
- Patrick Henry shouting "Give me liberty or give me death."

- A lion.

Symbols of the Left

- Gregory Peck arguing, "All men are created equal." in To Kill a Mocking Bird
- Martin Luther King saying "I've got a dream".
- Abraham Lincoln saying "A house divided against itself cannot stand."
- A bonobo monkey.

Religious Orientation

The religious orientation of the conservative-right is toward Jehovah and Allah. The religious orientation of the liberal-left is toward Jesus and Buddha.

Economics

Positions on economic issues can also be grouped along fairness -vs- freedom lines.

The conservative-right believes in the freedom of individuals to pursue wealth without interference from central authority. The inequalities that result when stronger, smarter, more aggressive people rise to the top of the economic heap are to be tolerated. Unfettered capitalism - without external controls is the most efficient system for managing goods and services.

The liberal-left believes that unfettered capitalism will result in an unfair concentration of wealth in the hands of the few. It also believes that in a managed economy where everyone gains wealth, the total wealth of the system increases. Even though the slices of the pie might be remain unequally divided, the total size of the pie gets bigger.

Capitalism, even when managed, seems to be a system of emergent rules - where the order springs from the system itself.

Sources

Following are sources checked (and double checked and triple checked) for this book...

- http://www.pbs.org/healthcarecrisis/history.htm
- http://www.randomhistory.com/2009/03/31_health-insurance.html
- http://library.thinkquest.org/15569/h
- http://library.thinkquest.org/15569/hist-4.html ist-4.html
- http://www.yalemedlaw.com/2009/11/the-history-of-medical-insurance-in-the-united-states/
- http://money.cnn.com/2012/07/12/news/economy/health-care-costs/index.htm
- http://www.kaiseredu.org/issue-modules/us-health-care-costs/background-brief.aspx
- http://www.ncbi.nlm.nih.gov/pmc/articles/PMC2517971/
- http://en.wikipedia.org/wiki/Fee-for-service
- http://www.ncbi.nlm.nih.gov/pubmed/21769757
- http://people.howstuffworks.com/why-pursue-health-reform3.htm
- http://www.pbs.org/newshour/rundown/2012/10/health-costs-how-the-us-compares-with-other-countries.html
- http://www.healthaffairs.org/healthpolicybriefs/brief.php?brief_id=78
- "The Healing of America" by TR Reid
- "Sick" by Jonathan Cohn
- "Free Lunch" by David Cay Johnston

www.ingramcontent.com/pod-product-compliance
Lightning Source LLC
Chambersburg PA
CBHW021911170526
45157CB00005B/2049